Tucholsky Wagner   Zola   Scott   Schlegel
Turgenev   Wallace   Sydow   Freud
                                 Fonatne
         Twain   Walther von der Vogelweide Fouqué   Friedrich II. von Preußen
                  Weber   Freiligrath
                                    Kant   Ernst   Frey
Fechner Fichte Weiße Rose von Fallersleben   Richthofen   Frommel
                                       Hölderlin
         Engels   Fielding   Eichendorff   Tacitus   Dumas
    Fehrs   Faber   Flaubert
                                  Eliasberg   Ebner Eschenbach
              Maximilian I. von Habsburg   Fock   Zweig
Feuerbach   Eliot   Vergil
                     Ewald
           Goethe   Elisabeth von Österreich   London
Mendelssohn Balzac   Shakespeare   Ganghofer
                      Lichtenberg   Rathenau   Dostojewski
         Trackl   Stevenson   Doyle   Gjellerup
Mommsen   Tolstoi Hambruch
         Thoma   Lenz   Droste-Hülshoff
                         von Arnim   Hanrieder
Dach   Verne   Hägele   Hauff   Humboldt
         Reuter
   Karrillon   Rousseau   Hagen   Hauptmann   Gautier
                 Garschin   Baudelaire
         Damaschke   Defoe   Hebbel
                        Descartes   Hegel   Kussmaul Herder
Wolfram von Eschenbach   Dickens   Schopenhauer   Rilke   George
            Darwin   Melville   Grimm Jerome   Bebel
   Bronner   Proust
       Campe   Horváth   Aristoteles   Federer
Bismarck   Vigny   Barlach   Voltaire   Herodot
                    Gengenbach   Heine
   Storm   Casanova   Tersteegen   Gilm   Grillparzer   Georgy
                 Chamberlain Lessing   Langbein   Gryphius
Brentano   Lafontaine
Strachwitz   Claudius   Schiller Kralik Iffland Sokrates
         Katharina II. von Rußland   Bellamy   Schilling
                                  Gerstäcker   Raabe   Gibbon   Tschechow
   Löns   Hesse   Hoffmann   Gogol   Wilde   Gleim   Vulpius
      Luther   Heym   Hofmannsthal   Klee   Hölty Morgenstern   Goedicke
           Roth   Heyse   Klopstock   Kleist
Luxemburg   Puschkin Homer   Mörike
              La Roche   Horaz   Musil
   Machiavelli   Kierkegaard   Kraft   Kraus
  Navarra Aurel   Musset   Moltke
                         Lamprecht Kind   Kirchhoff   Hugo
        Nestroy Marie de France
    Nietzsche   Nansen   Laotse   Ipsen   Liebknecht
                 Marx   Ringelnatz
      von Ossietzky   Lassalle   Gorki   Klett   Leibniz
               May   Irving
                      vom Stein   Lawrence
         Petalozzi   Knigge
              Platon
                   Pückler Michelangelo   Kock   Kafka
         Sachs   Poe   Liebermann
                                                    Korolenko
              de Sade Praetorius   Mistral   Zetkin

The publishing house tredition has created the series **TREDITION CLASSICS**. It contains classical literature works from over two thousand years. Most of these titles have been out of print and off the bookstore shelves for decades.

The book series is intended to preserve the cultural legacy and to promote the timeless works of classical literature. As a reader of a **TREDITION CLASSICS** book, the reader supports the mission to save many of the amazing works of world literature from oblivion.

The symbol of **TREDITION CLASSICS** is Johannes Gutenberg (1400 – 1468), the inventor of movable type printing.

With the series, tredition intends to make thousands of international literature classics available in printed format again – worldwide.

All books are available at book retailers worldwide in paperback and in hardcover. For more information please visit: www.tredition.com

tredition was established in 2006 by Sandra Latusseck and Soenke Schulz. Based in Hamburg, Germany, tredition offers publishing solutions to authors and publishing houses, combined with worldwide distribution of printed and digital book content. tredition is uniquely positioned to enable authors and publishing houses to create books on their own terms and without conventional manufacturing risks.

For more information please visit: www.tredition.com

Opportunities in Engineering

Charles M. (Charles Marcus) Horton

Imprint

This book is part of the TREDITION CLASSICS series.

Author: Charles M. (Charles Marcus) Horton
Cover design: toepferschumann, Berlin (Germany)

Publisher: tredition GmbH, Hamburg (Germany)
ISBN: 978-3-8491-8451-3

www.tredition.com
www.tredition.de

Copyright:
The content of this book is sourced from the public domain.

The intention of the TREDITION CLASSICS series is to make world literature in the public domain available in printed format. Literary enthusiasts and organizations worldwide have scanned and digitally edited the original texts. tredition has subsequently formatted and redesigned the content into a modern reading layout. Therefore, we cannot guarantee the exact reproduction of the original format of a particular historic edition. Please also note that no modifications have been made to the spelling, therefore it may differ from the orthography used today.

CONTENTS

CHAP.

I.	Engineering and the Engineer
II.	Engineering Opportunities
III.	The Engineering Type
IV.	The Four Major Branches
V.	Making a Choice
VI.	Qualifying for Promotion
VII.	The Consulting Engineer
VIII.	The Engineer in Civic Affairs
IX.	Code of Ethics
X.	Future of the Engineer
XI.	What Constitutes Engineering Success
XII.	The Personal Side

OPPORTUNITIES IN ENGINEERING

I

ENGINEERING AND THE ENGINEER

Several years ago, at the regular annual meeting of one of the major engineering societies, the president of the society, in the formal address with which he opened the meeting, gave expression to a thought so startling that the few laymen who were seated in the auditorium fairly gasped. What the president said in effect was that, since engineers had got the world into war, it was the duty of engineers to get the world out of war. As a thought, it probably reflected the secret opinion of every engineer present, for, however innocent of intended wrong-doing engineers assuredly are as a group in their work of scientific investigation and development, the statement that engineers 2 were responsible for the conflict then raging in Europe was absolute truth.

I mention this merely to bring to the reader's attention the tremendous power which engineers wield in world affairs.

The profession of engineering—which, by the way, is merely the adapting of discoveries in science and art to the uses of mankind—is a peculiarly isolated one. But very little is known about it among those outside of the profession. Laymen know something about law, a little about medicine, quite a lot—nowadays—about metaphysics. But laymen know nothing about engineering. Indeed, a source of common amusement among engineers is the peculiar fact that the average layman cannot differentiate between the man who runs a locomotive and the man who designs a locomotive. In ordinary parlance both are called engineers. Yet there is a difference between them—a difference as between day and night. For one merely operates the results of the creative genius of the other. This almost universal ignorance as to what constitutes an engineer serves to show to what broad extent the profession of engineering is isolated.

Yet it is a wonderful profession. I say this with due regard for all other professions. 3 For one cannot but ponder the fact that, if engineers started the greatest war the world has ever known—and engineers as a body freely admit that if they did not start it they at least made it possible—they also stopped it, thereby proving themselves possessed of a power greater than that of any other class of professional men—diplomats and lawyers and divinities not excepted.

That engineering is a force fraught with stupendous possibilities, therefore, nobody can very well deny. That it is a force generally exercised for good—despite the World War—I myself, as an engineer, can truly testify. With some fifteen years spent on the creative end of the work—the drafting and designing end—I have yet to see, with but two or three rare exceptions, the genius of engineers turned into any but noble channels.

Thus, engineering is not only a wonderful profession, with the activities of its followers of utmost importance, but also it is a profession the individual work of whose pioneers, from Watt to Westinghouse and from Eiffel to Edison, has been epoch-making.

For when James Watt, clock-repairer, tinker, being called into a certain small laboratory in England more than a century ago to 4 make a few minor repairs on a new design of steam-engine, discovered, while at work on this crude unit deriving its motion from expanded steam and the alternate workings of a lever actuated by a weight, the value of superheated steam for power purposes, and later embodied the idea in a steam-engine of his own, Watt set the civilized world forward into an era so full of promise and discovery that even we who are living to-day, despite the wonderful progress already made in mechanics as represented among other things in the high-speed engine, the dynamo, the airplane, are witnessing but the barest of beginnings.

Likewise, when George Westinghouse, inventor of the airbrake, having finally persuaded the directors of the Pennsylvania Railroad, after many futile attempts in other directions, to grant him an opportunity to try out his invention, and, trying it out—on a string of cars near Harrisburg—ably demonstrated its practicability as a device for stopping trains and preventing accidents, he also—as had

Watt before him—set the civilized world forward into an era full of promise and discovery as yet but barely entered upon, even with the remarkable progress already made in industry alone in the matter of regard for the safety of human life—Westinghouse's own particular blazed trail through the forest of human ignorance this same airbrake.

So with other pioneers—with Eiffel, in the field of tower construction; with Edison, in the field of electricity; with the Wright brothers, in the field of aerial navigation; With Simon Lake, inventor of the submarine boat. All were pioneers; all set the civilized world forward; all—though this perhaps is irrelevant, yet it will serve to reveal the type of men these pioneers were and are—all overcame great obstacles—Lake not the least among them.

Told that he was visionary, when Lake explained, as he did in his effort to enlist capital with which to build his first submarine boat, that he could safely submerge his invention and steer it about on the bed of the ocean as readily as a man can steer an automobile about the streets of a city, that while submerged he could step out of the boat through a trap-door without flooding the boat, by the simple process of maintaining a greater air pressure inside than the pressure of the water outside—Simon Lake, discouraged on every hand, finally decided to build a boat himself, and did build one, with his own hands—a boat fourteen feet long and constructed of rough pine timbers painted with coal-tar—in Atlantic Highlands, New Jersey. With this boat Lake demonstrated to a skeptical world for all time that he was neither a visionary nor a dreamer, but a practical doer among men—an engineer.

Of such stuff, then, were, and are, engineers made. Whether they realized it or not, whether the world at large realized it or not, each represented a noble calling, each was a professional man, each was chiseling his name for all time into the granite foundations of a wonderful profession even yet only in the building—engineering. Their name is legion, too, and their names will last because of the fact that their work, remaining as it does after them equally with the work of followers of the finest of the fine arts, is known to mankind as a benefit to mankind. Known by their works, the list extends back to the very dawn of history.

For it was men of this calling, the calling of engineers, who in the early days wrought for purposes of warfare — warfare then being the major industry — the javelin, the spear, the helmet, the coat of mail, the plate of armor, the slingshot; just as their later brothers, for a like purpose, conceived and devised the throwing of mustard gas, the two-ton explosive, the aerial bomb, the mortar shell, the hand-grenade — for the protection, false and true, of the home. For the upbuilding of the home, for the continuance of the home, men of this calling also it was who conceived and shaped, among other things, the cook-stove, the chimney, the wheel, the steam-engine, the spinning-jenny, the suspension-bridge, the bedspring-oh, boy! — the bicycle, the sandblast, the automobile, the airplane, the wireless.

Thus it will be seen that engineering is a distinctive and important profession. To some even it is the topmost of all professions. However true that may or may not be to-day, certain it is that some day it will be true, for the reason that engineers serve humanity at every practical turn. Engineers make life easier to live — easier in the living; their work is strictly constructive, sharply exact; the results positive. Not a profession outside of the engineering profession but that has its moments of wabbling and indecision — of faltering on the part of practitioners between the true and the untrue. Engineering knows no such weakness. Two and two make four. Engineers know that. Knowing it, and knowing also the unnumbered possible manifoldings of this fundamental truism, engineers can, and do, approach a problem with a certainty of conviction and a confidence in the powers of their working-tools nowhere permitted men outside the profession.

II

ENGINEERING OPPORTUNITIES

The writer can best illustrate the opportunities for young men which exist in engineering by a little story. The story is true in every particular. Nor is the case itself exceptional. Men occupying high places everywhere in engineering, did they but tell their story, would repeat in substance what is set forth below. More than any other profession to-day, engineering holds out opportunities for young men possessing the requisite "will to success" and the physical stamina necessary to carry them forward to the goal. Opportunities in any walk of life are not all dead—not all in the past. A young man to-day can go as far as he wills. He can go farther on less capital invested in engineering than in any other profession—that's all.

The young man's name was Smith. He was one of seven children—not the seventh son, either—in a poor family. At the age of sixteen he went to work in overalls on a section 10 of railroad as a helper—outdoor, rough work. At seventeen he was transferred to the roundhouse; at nineteen he apprenticed himself to the machinist trade. Engineering? He did not know what it was, really. Merely he saw his way clear to earning a livelihood and went after it. He was miserably educated. His knowledge of mathematics embraced arithmetic up to fractions, at which point it faded off into blissful "nothingness"—as our New-Thoughtists say. But he had an inquiring mind and a proper will to succeed. While serving his three years in the shop he bought a course in a correspondence school and studied nights, taking up, among other things, the subject of mechanical drafting. When twenty-two years of age he applied for, and got, a position as draftsman in a small company developing a motorcycle. He was well on his way upward.

He spent a year with this company. He learned much of value to him not only about mathematics, but about engineering as a whole as well. One day he decided that the field was restricted—at least, too much so for him—and he left and went with a Westinghouse organization in Pittsburgh. His salary was in the neighborhood of a hundred and ten dollars a month. He remained with the 11 company two years as a designer, and then, having saved up sufficient funds to meet his needs, went to college, taking special work — physics and chemistry and mathematics. He remained in school two years. When he came out, instead of returning to the drafting-room and the theoretical end of the work, he donned overalls once more and went to work in the shop as an erecting man. Two years afterward he was chief operating engineer in a small cement-plant in the Southwest, his salary being three thousand dollars a year. A year of this and he returned East, at a salary of four thousand dollars a year, as operating engineer of a larger plant. Then came a better offer, with one of the largest, if not the very largest, steel-plants in the country, as superintendent of power, at a salary of five thousand dollars a year. When the war broke out, or rather when this country became involved in the war, my friend Smith, at a salary of ten thousand dollars a year, became associated with a staff of engineers brought together into a corporation manufacturing shells. And all before he was barely in his thirties!

A young man still, what lies ahead of him can readily be surmised. Smith will follow engineering on salary until he is probably forty, when he will enter upon a consulting practice, and at fifty retire with sufficient money to keep him in comfort the remainder of his days. Nor will he be an exception, as I have stated in the opening paragraph. The profession is crowded with men who have worked up from equally humble beginnings. Indeed, one of the foremost efficiency engineers in the country to-day began as an apprentice in a foundry, while another, fully as well known in efficiency work, began life in the United States navy as a machinist's mate. Automobile engineers, whose names, many of them, are household words, in particular have gone big in the profession and from very obscure beginnings. It is not stretching the obvious to say that the majority of these men, had they entered upon any other work, would never have been heard from nor have attained to their present wealth and affluence. Smith was just one of many in a profession offering liberal opportunities. The opportunities still exist and in just as large a proportion as they ever existed. It remains but for the young man to decide. The profession itself, almost, will take care of him afterward.

However, not all of our engineers have gone upward by the overalls route. Nor is it at all necessary to do this in order to attain to success. The high-school graduate, entering a college of engineering, has an equal chance. Some maintain that he has a better chance. Certain it is that he is better qualified to cope with the heavier theoretical problems which come up every day in the average engineer's work. There is a place for him, side by side with the practical man, and his knowledge will be everywhere respected and sought. But a combination of the theoretical and the practical, as has frequently been declared, makes for the complete engineer. Some get the practical side first and the theoretical side later; some get the theoretical side first and the practical side later. It matters little—save only that he who gets the practical side first is earning his way while getting it, while the man who goes to college is in the majority of cases being supported from outside sources while getting what he wants. But in the end it balances. Merely, the "full" engineer must have both. Having both, he has, literally, the world within his grasp. For engineering is—to repeat—the adapting of discoveries in science and art to the uses of mankind. And both art and science reflect and are drawn from Mother Nature.

14

There is still a great scarcity of engineers. All branches feel the need—civil, mechanical, mining, chemical, automotive, electrical—the call goes out. It is a call just now, owing to the vast reconstruction period confronting the world, lifted in strident voice. Engineers everywhere are needed, which in part accounts for the liberal salaries offered for experienced men. The demand greatly exceeds the supply, and gives promise of exceeding it for a number of years to come. All manufacturing-plants, all mining enterprises, of which of both there are thousands upon thousands, utilize each from one to many hundreds of engineers. Some plants make use of three or four different kinds—mechanical, civil, electrical, industrial—some only one. But not a plant of any size but that has need for at least one engineer, and engineers are scarce. Therefore opportunities are ample.

To the young man seeking a profession, provided he be of a certain type—possessed of certain inherent qualities, the nature of which I shall set forth in the following chapter—engineering offers satisfactory money returns and—more satisfactory still—a satisfactory life. The work is creative from beginning to end; it has to do frequently 15 with movement—always a source of delight to mankind; a source having its beginnings in earliest infancy, and it is essentially a work of service. To build a bridge, to design an automatic machine, to locate and bring to the surface earth's wealth in minerals—surely this is service of a most gratifying kind.

And it pays. The arts rarely pay; science always pays. And engineering being a science, a science in the pursuit of which also man is offered opportunities for the exercise of his creative instincts, like art, is therefore doubly gratifying as a life's work. I know—and it will bear repeating—no other profession that holds so much of bigness and of fullness of life generally. Engineers themselves reflect it. Usually robust, always active, generally optimistic, engineers as a group swing through life—and have swung through life from the beginnings of the profession—without thought of publicity, for instance, or need or desire for it. Their work alone engrossed their minds. It was enough—it is enough—and more. And that which is sufficient unto a man is Nirvana unto him—if he but knew it. Engineers seem to know it.

16

III

THE ENGINEERING TYPE

It is becoming more and more an accepted fact that engineers, or physicians, or lawyers—like our poets—are born and not made. I believe this to be true. Educators generally are thinking seriously along these lines, with the result that vocational advisers are springing up, especially in industrial circles, to establish eventually yet another profession. Instinct leads young men to enter upon certain callings, unless turned off by misguided parents or guardians, and as a general thing the hunch works out successfully. Philosophers from time immemorial, including Plato and Emerson, have written of this still, small voice within, and have urged that it be heeded. The thing is instinct—cumulative yearnings within man of thousands of his ancestors—and to disobey it is to fling defiance at Nature herself. Personally, I believe that when this law becomes more generally understood there will be fewer 17 failures decorating park benches in our cities and cracker-boxes in our country stores.

The profession of engineering, therefore, has its type. You may be of this type or you may not. The type is quite pronounced, however, and you need not go wrong in your decision. All professions and all trades have their types. Steel-workers—those fearless young men who balance skilfully on a girder, frequently hundreds of feet in the air—are not to be mistaken. Rough, rugged, gray-eyed; with frames close-knit and usually squat; generous with money, and unconcerned as to the future; living each day regardless of the next, and *living* it—steel-workers are as distinct from the clerical type—slender, tall, a bit self-conscious, fearful of themselves and of the future—I say, the steel-worker is as different from the clerical worker as the circus-driver is from the cleric. Their work marks them for its own, if a man lack it upon entering the work, just as the school-room marks the teacher in time for its own. The thing is not to be mistaken.

The successful engineer must be possessed of a certain fondness for figures. The subject of mathematics must interest him. He must like to figure, to use a colloquialism, 18 and his fondness for it must be genuine, almost an absorption. It must reveal itself to him at an early age, too, as early as his grammar-school days, for then it will be known as genuinely a part of him, and the outcropping of seeds correctly sown by his ancestors. Having this fondness for mathematics, which may be termed otherwise as a curiosity to make concrete ends meet—the working out of puzzles is one evidence of the gift—the young man is well armed for a successful career in the profession. He will like mathematics for its own sake, and when, later, in college, and later still, in the active pursuit of his chosen work, he is confronted with a difficult problem covering strains or stress in a beam or lever or connecting-rod, he will attack it eagerly, instead of—as I have seen such problems attacked more than once—irritably and with marked mental effort.

The successful engineer must be a man who likes to shape things with his hands. He need not always do it, and probably will not after he has attained to recognition, save only as he supervises or makes the mechanical drawings—the picture—of the thing. But the itch must be present in the man. And, like the desire within him to figure, it must make itself manifest within him early in life. If a young man be of those who early like to crawl in under the family buzz-wagon; tinker there for half a day at a time; emerge in a thick coating of grease and dust and with joy in his eye—such a young man has the necessary qualifications for a successful engineer. He may never do this—as I say—in all his engineering career. But the yearning must be as much a part of him as his love for mathematics—so much so that all his engineering days he will feel something akin to envy for the machinist who works over a machine of the engineer's own devising—and it must be vitally a part of him. To illustrate:

When only twelve years old the author, in company with several playmates, decided one November day to build an ice-boat. From the numerous building operations going on in the neighborhood, in the light of the moon, he secured the necessary timbers, and from a neighbor's back yard—also in the light of the moon—he got a young sapling which served delightfully as a mainmast. With the needed materials all gathered, it suddenly struck him that a plan of some kind ought to be made of the proposed ice-boat, in order to guard against grave errors in construction. To think was to act with this bright youngster. He got him his mother's bread-board and a pencil and an ordinary school ruler, and with these made a drawing of the ice-boat as he thought the boat should be. Knowing nothing of mechanical drawing, and but very little of construction of any kind, he nevertheless devised a pretty fair-looking boat and not a bad working drawing. One of his playmates, whose father was something or other in a manufacturing-plant, showed the drawing to the family circle; with the result that the kid's father, laying a rule upon the drawing, pronounced it an accurate mechanical drawing, drawn to scale — which was one inch to the foot — and sent for the youthful designer, meaning me.

"What do you know about mechanical drawings?" he asked the bashful youngster, pointing to the drawing under discussion.

"I don't know nothing about it," replied the kid — meaning me again. "I just made it with a ruler."

"But how come you made it to scale? That drawing is a complete plan and elevation of an ice-boat, drawn accurately to scale." He looked thoughtful. "I don't understand it. You ought to take up with drafting, my boy, when you get a little older. 21 I never knew of a case like it. What does your father do?" he suddenly asked.

"He's an ice-dealer," [1] replied the discomfited boy. "I just made it — that's all. We need it, too, to go ahead." Turning to his playmate, "Come on out, Jack; the gang is waiting."

Which terminated the interview.

Yet the thing was the beginning of a career for the boy. The boat in time somehow got itself built and out upon the little river; but owing to the fact that its materials were stolen, the river failed to freeze over that winter, and for three winters following — not till the boat itself had fallen apart from disuse and lack of care — which points its own moral, as hinted at above. If you must build ice-boats, and you are a kid with mechanical yearnings, pay for the material that goes into the making of your product. But the thing — as I say — was the beginning of a career for the lad. In time, through the kindly office of his playmate's father, he became apprenticed in a drafting-room of a large manufacturing-plant — and the rest was easy. In his first year, on paper, he devised a steam-engine with novel arrangement of slide-valves, and thereafter for years designed

22 engines and machinery about the country, always quite successfully.

The successful engineer, while possessed of certain spiritual characteristics, must also — if I may be so bold as to say so — be possessed of certain physical characteristics. One of these is large, and what is known as capable, hands. Short, spatulate fingers, with a broad palm, appear to be a feature of the successful engineer. Of course, there are exceptions, as there are exceptions to every rule, but in the majority of cases which have come under the writer's observation the successful engineer has had hands of this shaping. He likewise has had wrists and arms to match with such hands, and — in the practical engineer — that is, the engineer whose particular gift is coping with ordinary problems of construction, as against the genius who blazes new trails, like Watt and Westinghouse and Edison and Marconi and the Wright brothers — a head whose contour was along the "well-shaped" lines. The so-called genius usually has an odd-shaped head, I've noticed, but for purposes of this book we shall confine ourselves to the average successful man in engineering.

Thus you have, roughly, the engineering type. I have sketched only the major characteristics. 23 The minor characteristics embrace many features. There is patience, for one — patience to labor long with difficulties; concentration, for another; application, for a third; certain student qualities, for yet a fourth. Many graduate engineers have gone off into other work immediately after leaving college because of a clearly defined dislike for detail in construction. The average successful engineer will be a man interested in the shaping of the details of his machine or bridge or plant. To many, details are irksome. If the young man who is reading this book knows that he dislikes a detail of any character whatsoever, unless he be possessed of the creative genius of a Westinghouse or an Edison, he would better take up with some other profession. For engineering, in the last analysis, is the manipulating of detailed parts into a perfect whole — whether it be a bridge or a machine or a plant.

FOOTNOTES:

[1] The boy's father always wanted to be a carpenter.

IV

THE FOUR MAJOR BRANCHES

The four major branches of engineering are civil, mechanical, electrical, and mining. I give them in the order of their acceptance among engineers. Each is separate from each of the others, and each is a profession in itself, and as distinctive from each of the others as is the allopathic from the homeopathic among men of medicine, though not with quite the same distinction. Whereas the several groups of physicians seek to relieve pain and correct disorder by way of diversified channels, the several groups of engineers each work in a field of endeavor actively apart from each of the other groups. Sometimes one group will lap over upon another group, in certain kinds of construction work, but even then the branches will hold sharply each to its own.

Civil engineering embraces, roughly, all work in the soil. The surveyor is a civil engineer. He constructs dams, builds viaducts, lays out railroads, and in the war, where he was known as a pioneer, he was responsible for all tunneling and trench projects, besides keeping the highways clear and the wire entanglements intact. Civil engineering is a profession which keeps its followers pretty well out in the open. A civil engineer will go long distances, and frequently must, in order to get to his work, and, having reached the scene of his labors, enters upon a rugged outdoor life in camp where he remains until the job is completed. The Panama Canal was a civil-engineering job — probably the largest of its kind ever undertaken — and its success, after failure on the part of another government, is a high tribute to the genius of our own civil engineers.

Mechanical engineering is a profession whose medium of endeavor lies in the metals. Mechanical engineers shape things out of iron or steel or brass or other metal compositions, and put these things into engines or machines for service. All machinery, whether it be printing-presses or automobiles or steam-engines, is the work of mechanical engineers, though in the matter of automobiles this has become a profession by itself, one of the minor branches known as automotive engineering. The mechanical engineer as a rule works within doors, just as the civil engineer works out of doors, and his work, consequently, is more confining. In the pursuit of his profession he spends much of his time supervising the design of mechanical units, and is the one man responsible for correct construction and security against fracture of the machine itself when in operation. Actually the mechanical engineer has more opportunities in his daily routine for the exercise of his creative faculties than has any one of the other kinds of engineers, for the simple reason that no two machines even for the same purpose—speaking of types, always—are exactly similar in construction. Two lathes of like size and scope, if manufactured by two separate organizations, will be different in their minor features, and each in some particular will be the work of a mechanical engineer whose ideas are at variance with those of the mechanical engineer who designed the other type. Engineers, like doctors, often disagree, which accounts for the many different types of machinery serving the same purpose which are found on the market.

Electrical engineering is, as its name implies, a profession embracing all construction whose basis is the electrical current. Any unit whatsoever, so long as it utilizes or eats 27 up or carries forward a current of electricity, is the work of electrical engineers. The profession is a comparatively recent one perforce, owing to the fact that but very little of a practical nature was known about electricity until a very few years ago. The wonderful progress in this field made within the past twenty years is one of the marvels of the engineering profession. Dynamos, motors, arc-lights, alternating current, the X-ray—these are a few of the things which followers of the profession have created for the uses of mankind. The field is yet practically unexplored, and offers to engineering students an outlet for their energies—provided they enter this branch of engineering—second to none of the other branches. A fascinating study, doubly so because of the fact that nothing is known about electricity itself—its effects only being understood—electrical engineering should appeal to the curious-minded as no other vocation can. It is a profession shrouded in mystery, and not the least mysterious of its recent developments is the wireless telegraph. What this one development alone holds for the future nobody can say. All sorts of inventions can be imagined, however, and among them I myself seem to see automobiles operated 28 from central stations—indeed, all mechanical movements so operated—to the end that individual engines in time will cease to be.

The profession of mining engineering, last of the major branches, embraces all work having to do with the locating and construction of mines—coal-mines, iron-mines, copper-mines, diamond-mines, gold-mines, and the like. Also it establishes the nature of the apparatus used, though more often than otherwise the mechanical engineer in this regard is consulted, since much of the machinery utilized in mining operations is the direct work of mechanical engineers. Screens and hoppers are mechanical devices the result of mechanical engineering genius; but the work of shoring up, done with timbers, and the work generally of supervision of all mine operations, rests solely with the mining man. The shaping of these timbers, though—the cutting of tenons, for instance—is the work, again, of the mechanical engineer; though the placing of these timbers, to revert back once more, is the work of the mining engineer.

There are many minor branches, and more are rapidly coming into prominence. Chemical engineering is one of the older minor branches; while industrial engineering—following closely upon automotive engineering—belongs properly with the more recent of the newcomers. Efficiency engineering is a branch which to-day is making a strong bid for recognition as a profession, although the work as yet, lacking, as it does, proper foundation in scientific truth, even though strongly humanitarian in its motives, has still to prove itself acceptable among the engineering groups. Structural engineering, on the contrary, "belongs." Its work consists of the design and layout of modern steel structures—this roughly—while the minor branch known as heating and ventilating engineering, as its name would indicate, deals with the proper heating and ventilating of buildings, and as a profession is closely allied with that of structural engineering. Out of these minor branches come yet other branches, more particularly groups, with each in the nature of a specialty, such as gas engineering, aircraft engineering, steam engineering, telephone engineering, and so on.

Students about to enter engineering colleges usually select one or another of the major branches and then after graduating begin to specialize. But infrequently Fate has much to do with this specialization, since after leaving college the average young engineer will turn to the nearest or most promising 30 vacancy offered him in his chosen field—a major branch—and in the work eventually become expert and a specialist. If it be a concern manufacturing steam-turbines, say, the young engineer in time becomes expert and a specialist in steam-turbines. So, too, with graduates in mining engineering, in electrical engineering, in civil engineering, although the opportunities for specialization in any of these latter branches are not so good as in the mechanical field. However, entering upon a certain kind of work, the student usually follows this work to the end of his days, which is probably what engineering schools expect. All strive to educate only in the principles of each of the major branches. The rest is up to the graduate, who is permitted, and generally does, the shaping of his own career afterward.

It is a feature of our democratic form of government—thanks be! Germany does—or did—the other thing. Germany made careers for her young men, instead of young men for careers, with the result that she also made machines out of them. America is a nation of individualists, which is what makes America what it is, and our schools and school systems are responsible.

31

V

MAKING A CHOICE

About to make a choice among the branches of engineering, the prospective student, unless he have a decided preference to start with, finds himself confronted with many difficulties. Engineering is engineering, whether it be mining or electrical or civil or mechanical, and this fact alone is not without its confusions. Yet if the young man decides for a mining career, say, the choice may take him, after graduating, off to South Africa, whereas if his choice lay in the electrical field he may never get any farther from home than the nearest electrical manufacturing plant in his town or state — and remain there for the duration of his life. This making of a choice is a momentous thing in a prospective engineer's life. It should be approached with all caution, and with due regard for the nature of the life he would lead after graduating from school. If he have a penchant for outdoor life, then the 32 choice, in a way, is easy. He should select mining or civil engineering as his particular vocation. If he be of those who prefer to remain more or less indoors in the practice of his profession, mechanical or electrical engineering should be his choice.

These are the major advantages or disadvantages, depending upon the point of view. The minor ones are not so easily stated. Speaking always for the young man without a decided preference, it is the writer's opinion that the prospective student should analyze his particular feelings in the matter and decide accordingly. Large projects may interest him more than smaller ones. In this regard, he will find greater satisfaction in following the profession dealing with large projects, which is, of course, the civil engineering profession — although mining, too, has its large ventures, which, however, do not "break" as frequently as they do in civil engineering. On the other hand, the young man may find himself attracted to the development of small propositions, such as adding-machines and typewriters and sewing-machines, and the like. Finding himself attracted to these no less important phases of engineering than the development of mines or the opening up of new country, the 33 young man can, of course, make no better choice than to enter the mechanical or the electrical field.

It all depends upon the point of view. Nor is there any hard-and-fast rule tying a man down to a single branch once he finds that he does not like it, or finds that he likes one of the other branches better, after he has given his chosen branch a trial in the years immediately following graduation. Not a few mining graduates drift over into straight civil work after leaving school, and, likewise, not a few in the electrical branches find themselves in time pursuing mechanical work. Fate here, as in the matter of specialization, works her hand. A prominent publisher of technical magazines in New York took the degree of Arts in Cornell in his younger days; and more writers of fiction than you can shake a stick at once labored over civil-engineering plans as their chosen career. Herbert Hoover is a mining man who best revealed his capabilities in the field of traffic management—if the work which he supervised in Belgium may be so termed. Certainly it had to do with getting materials from where they were plentiful to where they were scarce, which is roughly the work of the traffic manager.

34

And so it goes. The young man in this particular must decide for himself. Actually, there is more of mystery and fascination in the electrical field than in any of the other three branches, and to prospective students this may be not without its especial appeal. To others, the work of mining may possess its strong attraction, since this work takes its followers into strange places and among strange people frequently, where oftentimes the mining engineer must live cheek by elbow with the roughest of adventurers. To yet a third group, civil engineering, with its work of blazing new trails through an unknown country, and wild outdoor existence through forests and over mountains and across valleys—may have its strong attraction. While to a fourth group of prospective students the quiet career, as represented in that of mechanical engineering, always a more or less thoughtful, studious life, may hold out its inviting side. The mechanical engineer, like the electrical engineer, is a man who generally commutes, a man who comes and goes daily between office and home, doing his work at regular hours within the four walls of his office—a quiet, professional man. Such a life would appeal to the man of family rather more strongly than either of the outdoor 35 professional branches. Yet the prospective student must make his own choice.

To the young man who has no particular preference, and who would put it up to the writer as to just which branch to follow—the young man more or less in need—the writer unhesitatingly would advise mechanical engineering. It is the one branch offering the largest and quickest returns, and as a branch it fairly dominates all the other branches, for the reason that whereas the mechanical engineer can get along without the mining engineer or the civil engineer or the electrical engineer, neither the mining engineer nor the civil engineer nor the electrical engineer can always do without the services of the mechanical engineer. No other branch so overlaps the other branches as does mechanical engineering. The work of the mechanical engineer is seen in almost every piece of construction reared by the civil man, just as it is seen in every bit of construction work of the mining and the electrical engineers. At first glance this may not appear to be true, but a close analysis of different jobs will bring out the truth of this statement.

Thus mechanical engineering offers largest and quickest returns. It does this for another 36 reason. Because of this very overlapping upon the other three branches, for every position open in the electrical field, or the mining or the civil field, there are a dozen vacancies in the mechanical field. It cannot but be otherwise. Not one of the other branches but what has need at times for—as I have stated—a mechanical engineer. The casings and base-plates and supports of motors, for instance, while the motor itself—its windings and the like—is the work of the electrical engineer, are due to the designing genius of some mechanical man. Likewise, in the mining field, where shaking screens, to name only one of the many mechanical units necessary in mining operations, are an essential factor—units operated with pulleys and belts and cams and levers—all the province of the mechanical engineer—the mechanical man finds his uses. So in civil work, especially in dam construction where gates are necessary; and in chemical engineering—to drop into a minor branch—where tanks and vats and ovens and stirring paddles and the like are used. No matter in which branch a man may go, always he will find evidence of the presence some time of the mechanical engineer. The mechanical engineer dominates all the other branches, 37 as has been said before. He is given second place in the order of the branches merely because the civil engineer happened to be the first and oldest kind of engineer to be given recognition as a profession. This man made himself a professional man, just as did the early practitioners of medicine—concocters of herbs in the beginning.

The proper selection will depend upon the young man's predilections and tastes. If he selects wisely, following out his predilections and tastes with a degree of accuracy, he cannot go wrong. He cannot go far wrong even if he doesn't follow out his hunches, for the reason that he can always swing over into any one of the other branches whenever he sees fit to do so. The thing is done every day, and will continue to be done throughout all time. Merely, it would be well for the young man, of course, to select in the beginning that branch which most appeals to him, and to stick to it like glue. Success is certain to be his. For in no other walk of life are the rewards so sure and so ample and so immediately responsive as in the engineering professions. These—like the matter of his selection from among the four major branches—are solely a matter up to the individual.

38

VI

QUALIFYING FOR PROMOTION

Immediately upon graduating—indeed, often several months before graduating—the engineering student finds his first job awaiting him. Frequently he finds a number of first jobs awaiting him and must make a selection. For it is the custom with large manufacturing concerns to send out scouts in the early spring of each year to address the engineering student bodies, with the idea in mind of securing the services of as many graduates as the scouts can win over for their respective organizations through direct appeal. What is usually offered the coming graduate is a brief apprenticeship in the shop, at a living wage, with promise of as early and rapid promotion in the organization as the work of the apprentice himself will permit, or improves.

These offers are generally splendid opportunities. The graduate may learn much of a practical commercial nature which perforce has been denied him in his student days, and also, having entered upon this apprenticeship, he not only gets acquainted with production on a large scale, but he is brought into touch with what constitutes most recent acceptable practice as well. This, provided he be a mechanical or an electrical engineer. Graduates in civil and mining engineering, while offered positions from executives in these particular branches also, have no such large opportunities offered them. The work itself does not permit it. Yet in any of the branches there is never a scarcity of jobs open to graduates upon their leaving college.

To qualify for promotion in any work, but more especially in the professions, one must know one's business. That is a trite statement, but it will bear repeating. The young graduate at first will not know his business. His mind will be a chaos of theories based upon myriads of formulæ which cannot but confuse him in the early days, when he is most earnestly trying to apply one or more of them to the more or less petty tasks which will be assigned to him. All he can do under the circumstances—all anybody could do under the circumstances—is to wait patiently, the while doing the best he can. Problems have a way of working themselves out—the correct formula will present itself; its true application will become manifest—and thus the young engineer has learned something of a practical nature which need not forsake him throughout the remainder of his engineering career.

Engineers are especially tolerant of one another's mistakes and errors. They are much more so than medical men, for instance. In the field of medicine one must show by many practical cases wherein a certain treatment has proved effective before the fraternity at large will even give the practitioner a hearing. This is not so among engineers. Engineers turn to one another in difficulties with earnest desire to help if they can help; and when one of their number is in trouble in his efforts to solve a difficult problem the whole body will turn to him with friendly encouragement and advice, if the latter is wanted. The young graduate who is struggling with a problem come up in his daily work, if he will but make the fact known to the engineers on the job in association with him, will find himself surrounded by engineers every one of whom will be seriously concerned for him and anxious to render assistance.

So the young graduate need entertain no fears on the ground of possible errors when starting out. Merely he must go slow; take his own good time on a job; ask all the questions possible of his engineer neighbors. Frankness in engineering, as in any other walk of life, pays. The bluffer is not wanted. No man knows it all, and certainly no engineer knows all there is to know about his profession. Time was when this might have been true; but it isn't true today. The work of engineering research and development has become so complex that engineers are forced to specialize. The engineering graduate, entering upon his first job, will discover early that he, too, must specialize. This will not be difficult, owing to the fact that his engineering education has been general and designed to embrace in a liberal way all practice. Drawing, as he will, from this liberal source that which he finds necessary in the solving of his initial problems, he will find himself within a short time becoming, willy-nilly, a specialist.

In the earlier years there should be considerable study done after hours on the part of the graduate engineer. Because his education has been general in the field, and he now holds a position with a company manufacturing steam-turbines, say, he must "wise up," as the saying goes, on the subject of steam-turbines. It will do him no harm to 42 trace back to its source all progress made in the field of turbine engineering and construction. He will find no scarcity of books on the subject, and with every hour spent with these volumes he will become more valuable to the organization employing him. Likewise, if he find himself working for an electrical manufacturing concern, and himself a graduate in electrical engineering, if the product be only a single line, and so small a thing as spark-plugs, it will profit him greatly to read whatever has been printed on the subject of spark-plugs. So with the mining graduate in the matter of the different processes of recovering minerals; so with the civil graduate, especially in the concrete field of construction, which has made rapid strides in the past few years—the graduate should absorb as much as he can of the available works printed on the subject. Indeed, this is the profession of it, in that the practitioner must ever be alive and alert to what is being done and has been done from the beginning in his chosen line of endeavor.

Next must come fealty. The graduate on his first job must believe—and if he does not believe ought to change connections—that the product of his company is the best in the market. This need not necessarily be 43 true; but he must feel that it is true. For only in this way can he put the best that is in him into his work. Industry—and the engineer is the backbone of industry—is a hotbed of competition. Any organization needs all the enthusiasm it can get. Greatest enthusiasm of all must come from within its own circles. Lacking this enthusiasm within its own family, the organization as a whole suffers. The graduate must first of all supply enthusiasm to the source of his employment, because at first he can supply but very little else. He must be true to his trust in ways other than the mere doing of what he is told or producing what he is expected to produce. This attitude cannot but help him qualify for promotion, and rapidly. It is a very important factor in any engineer's advancement.

Then there is the matter of patience. The writer knows of no other qualification more fruitful of reward than patience. The word control is frequently used in this regard—self-control. Its other name, however, is patience—the thing that gives a man to try and try again until he succeeds. Engineering is a difficult profession, though not more difficult than other professions, and in the average engineer's working-day many things occur which, if he be not possessed of 44 infinite patience, will serve to try him to a considerable degree. Patience with those below him—patience with those above him—patience with himself—these are all necessary and will prove helpful to him in reaching the top. He must accept the petty tasks with a cheerfulness no less apparent than he accepts the more important ones. He must present his own ideas to his superiors with a degree of caution which, where the ideas are rejected, will yet permit him to withdraw within himself without giving the impression of being peeved. For engineering is above all other things the interchange of ideas among men having an equal training but a vastly different quality of experience. Men of diverse experience thus drawn together make for a balanced engineering staff, and a balanced engineering staff makes for a well-organized whole. The young engineer must conduct himself in such a way that his superiors will like him for what he is, as indicated by his personality, rather than for what he knows or does in his daily work.

To sum up, then, the young engineer, having entered upon his first job, must do three or four things in order quickly to qualify for promotion. He first of all must spend time in study after his day's work is done—absorb 45 all information having to do with the company's own product; hold himself ever alert to the company's own methods of production; watch for an opportunity whereby this production may be improved upon or the methods of production themselves improved upon. The young engineer must proceed slowly in everything he undertakes; when brought to a halt through difficulties he should instantly appeal to one or another of his associates or superiors; he must be absolutely frank in all his dealings with these associates and superiors. In this regard, also, it might be said that the young graduate, following a habit become almost second nature with him in his school-days, must keep a note-book covering his activities throughout each working-day, a book wherein he will jot down everything of value to him which comes up in the day's work. Such books often form the basis of complete textbooks in after years, and, indeed, are acknowledged to be the foundation of more than one recognized authority. Though in this regard, further, such a practice is sometimes discouraged in some organizations, since it is apparent that these note-books often contain facts which the organization does not wish to have made public, 46 being, as these notes often are, in the nature of trade secrets. However, the student with a conscience will effectively guard the secrets of his employer as contained in his note-book, holding its contents for his own use in furthering the interests of the company which employs him.

And finally—in the matter of personality—patience and regard for the foibles of others will go far toward advancing the young engineer toward success. He must never forget in his earlier years that he is embryonic in the profession; that the profession is a difficult one and with many ramifications; that if he was able to live through three normal lives he would yet know only a very little of what there is to know about his chosen work. Thus he will conduct himself in a manner designed to win the interest and affection of men who are superior to him. Life to-day consists more than ever of service, and no man can go the path alone. Service—assistance one to another—makes up the sum total of life. No engineering graduate—no young man in any walk of life—can progress far without assistance, however brilliant as a student and capable as a man he may be. If he will but bear this last in mind—this and the other even more important 47 truth, that as a man gives so shall he receive—that a dollar spent in charity means two dollars in the bank—I mean that exactly—then the heights themselves will beckon to him at an early age.

"Early to bed and early to rise"; "take care of the pennies and the dollars will take care of themselves"; "a bird in the hand is worth two in the bush"—we don't need—the engineering graduate does not need—that form of admonition. It means nothing and is false. What alone counts for success is a considerable regard for the rights and privileges of others, the unfortunate as well as the fortunate. Greed never brought success that was lasting to any one, and certainly it breeds unhappiness. Engineering is a work of service—service to others—and to the graduate who "gets" this truism will come all things of this life, not the least of which will be material rewards.

48

VII

THE CONSULTING ENGINEER

The consulting engineer represents the pinnacle, as it were, of professional success. The inventor is something else—a wilding in the profession—and as such cannot be considered in a paper of this kind, save only as to say that he is the presiding genius among engineers, the Shakespeare or Milton among his kind, a man whose path to the heights is nowhere known of men. The consulting engineer, on the contrary, representing, as he does, the zenith of slowly attained power in some certain branch of engineering, a vantage—point open freely to all, is the embodiment of the goal toward which all graduates should strive. The consulting engineer has perfected himself in his chosen field; he has become an authority in his branch of engineering; his word is accepted as final in court and privy council. Having gained to this enviable position only after prolonged study and protracted and wide 49 experience in his particular specialty, the consulting engineer has well earned whatever accrues to him in the way, among other things, of generous fees for his services.

Still, there are consulting engineers who have become so through accident. The writer personally knows a consulting engineer who was following a general engineering practice when called upon one day to advise a group of capitalists in the matter of a garbage-disposal plant of new design for a large mid-Western city. His services were sought not because he was a garbage expert, but rather because he was expert in intricate pipe layouts and the like. However, once he got his hand into garbage disposition on a large scale, he remained in this branch of engineering, eventually traveling about the country supervising the design of similar plants whose object was the economical disposal of municipal refuse. Practically alone in the field, his writings soon became accepted as authoritative, and yet the whole thing began with that first call, quite by chance, in a matter foreign to the subject. Like other professional men, engineers never know when the heavens will open for their particular benefit.

Yet these cases are rare. The average consulting 50 engineer is a man who has won to pre-eminence only through protracted study and hard work in one line. He is a specialist with a high reputation for accuracy and skill in that line. The basis of this skill, of course, lies in a broad general engineering experience, upon which is built a peculiar knowledge of a certain, and not infrequently isolated, branch of engineering. Heating and ventilating engineers are but specialists grown to such large numbers as to form a definite branch of engineering. Likewise, automotive engineers are men who have specialized through long years in this branch. The man who knows more about building dredges, say, than any other man among his engineering brothers is a man who will be most frequently sought by industrial powers feeling the need for a dredge, just as a man suffering eye-strain will seek out the best specialist known to the medical fraternity. He goes to the one acknowledged authority in this line, and in doing so but follows a sane inner dictation.

And that is consulting work. The individual of money who would launch into manufacturing, knowing nothing of manufacturing, will, after deciding as to which branch of manufacturing he wishes to follow, 51 enlist the services of a consulting engineer big by reputation in this branch. The capitalist may wish to enter the paper-manufacturing field. Straightway he will put himself in touch with a consulting engineer whose specialty is paper-manufacturing plants, and, having informed this man as to the amount of money he is willing to spend on the venture, together with the location where he wishes, within certain prescribed limitations, to have his plant stand, may withdraw from the thing, if he choose, until the plant is built and in operation. The consulting engineer has done the rest. He has gone out upon location, seeking sites with an eye to economy both of power and transportation; he has supervised the design of the plant and the location in the plant of the necessary machinery; has enlisted the service of a builder whose task it is to follow these plans from foundation to roof in the work of actual construction. For this work the consulting engineer receives a fee, usually based upon a percentage of the cost, and then turns to other clients—waiting in his outer office—who would enlist his services in a similar capacity.

The consulting engineer has other sources of revenue. Like the lawyer, he is frequently 52 retained by traction and lighting interests to guard the rights of these interests, service for which he receives payment by the year. His testimony is valued in matters of litigation, sometimes patent infringements, sometimes municipal warfare between corporations, but always of a highly specialized nature. He is an authority, and when I have said that I have said all. His retainer fees are large; his work is exact; he is a man looked up to by those in the profession following a general practice. He has his office, and retains a staff of engineers, usually young engineers just out of college, who, like himself at one time, are on their way upward in the game. He is rarely a young man; generally is a man of wide reading; is a man respected in his community not for what he knows as an engineer, but for the standard of living which he is able to set by virtue of his income. Besides the sources of revenue which are his, and as I have set forth above, he is sought by technical editors to contribute to magazines powerful in his field, and this is a pleasurable source of income to any man in any walk of life. The consulting engineer is a man to be admired and emulated by all engineering students.

As to the time in life when an engineer 53 feels qualified to enter upon consulting work, that is something which must come to him from within. Usually the engineer knows that he has become a factor in his chosen branch or specialty when he finds himself becoming more and more sought in an advisory capacity among his fellows. He can judge that he has become an authority in his work by the simple process of comparing himself and his work with others and the work of these others in the field. If he finds that he is designing a better plant or automatic machine, or more economically operated mine or more serviceable lighting station than his neighbor, and, together with this knowledge, perceives also that capitalists are beating a deeper path to his door than to the doors of his competitors—to warp an Emersonian phrase—then the handwriting on the wall should be clear to him—to quote the Bible. Having sufficient capital to carry him through a year or two of personal venturing in the consulting field, he will open an office and insert his professional card in the journals in his field—and fly to it. If he be a man of righteous parts, he will succeed as a consulting engineer—and can go no higher in the profession.

The game is certainly worth the candle.

54

VIII

THE ENGINEER IN CIVIC AFFAIRS

Much has been written of late of the engineer as a citizen — of his civic responsibilities, of his relation to legislation, to administration, to public opinion, and the like. It is timely writing. The engineer is about due for active participation in civic affairs other than a yearly visit to the polls to register his vote. He has not done much more than this since his inception. His work alone has sufficed, for him, at least, though the time is past when he can bury himself in his professional work and, in the vernacular, get away with it. Men of the stamp of Herbert Hoover have demonstrated the very great need for men of scientific training in public affairs. Such places heretofore have been filled with business men and lawyers. These men served and served well. But since administration of public affairs to-day is largely a matter of formulation and execution of engineering projects, it is assuredly 55 the duty of engineers to take an active part in these public affairs.

Exact knowledge, which in a manner of speaking is synonymous with the engineer, is needed in high places in our nation. Men of technical education and training have demonstrated their fitness as servants of the people in the few instances where such men have taken over the reins of administration in certain specified branches of our government. Trained to think in terms of figures and the relation of these figures to life, engineers readily perceive the true and the untrue in matters of legislation and administration, though as a body they have never exerted themselves to an expression of their opinions on matters coming properly under the head of public opinion. Engineers have felt that they have not had the time. Or, having the time, that the public at large, chiefly owing to the engineer's self-imposed isolation, would not understand a voice from this direction, and so engineers have kept silent. The day has arrived, however, when this silence on the part of engineers must be broken.

The World War has been an awakening in this as in other directions. Lawyers and politicians have successfully dominated our 56 government from its beginning, with a single beautiful exception in

George Washington at one end and another admirable exception in Woodrow Wilson at the other. Washington was a civil engineer, and Wilson, while trained as a lawyer, was an educator. In between these two men there may have fallen a scattering of others who were not lawyers or politicians; the writer is not sure. Of one thing he is sure, however, and that is that engineers in the future will dominate politics to the betterment of the nation as a whole. For engineers are idealists—otherwise they would never have entered upon an engineering career—and idealism has come, as it were, into its own again. The man of vision of a wholesome aspect, the man who can so completely forget himself in his work of service as to engage in tasks whose merits nobody save himself and those pursuing like tasks can or will understand—which is pre-eminently the engineer—is the one man best fitted to administrate in public affairs. More important still than this statement is the fact that the world at large is beginning to realize the truth of it. Engineers as a body stand poised upon the rim of big things. Nor will they as a body stoop to the petty in politics, once they are fairly well launched 57 in active participation of civic affairs. Neither their training nor their outlook, based upon their training, will permit it. For engineers, more than any other group of professional men, are given to "see true." And seeing true, being, as it is, the essence of a full life, is what is needed in our public administrators.

Engineers in the past who have become more or less prominent in the public eye—and there are some who have—have demonstrated their ability to see things as they are. Westinghouse was the first man in this country to foresee the coming of the half-holiday Saturday as an innovation that promised general adoption. He granted it to all his employees at a time when lesser industrial captains believed him to be at least "queer." Ford set the pace for a minimum rate of five dollars a day in his plant, and lesser captains still frown upon him for having perpetrated this "evil." Edison, among other things, has told of the importance of loose clothing—loose shoes and collars and hats—to a man who would enjoy good health. The list is not long, but the insight of those who form this short list cannot but be recognized. What these men have said and done concerning matters freely apart from the subject of 58 engineering reveals them as members of a fraternity well qualified to lead public opinion rather than to follow it, as has been the province of engineers in the past. Each when he has spoken or entered upon action having the public welfare in mind has pronounced or demonstrated a truth which fairly crackled with sanity.

Engineers belong in civic affairs. The world of humanity needs men of their stamp in high places. Humanity needs men in control of state and national affairs who would hold the interests of humanity sacred. Engineers are such men. Not that engineers more than any other professional men are sprouting wings—not that. But engineers do see things in their true light—cannot see them in any other light than the one imposed by the law of mathematics, which is that two and two make four, never five or three—and this involuntarily would admit of decisions and grant graces from the point of view of absolute truth, which is, of course, the point of view of humanity—the greatest good for the greatest number. With such men occupying high places in the nation's affairs, the world of men and mankind would leap forward ethically and spiritually at a pace in keeping with the pace at which civilization has progressed under the impetus of engineering thought since the days of Watt. Nobody can deny *that* progress. Nobody could well deny the fact that ethical progress under engineering guidance would be equally great.

I hold a brief for engineers, of course. Engineering has been my major work for twenty years and more. It has been my privilege to associate intimately with two men—yea, three—possessed of great engineering ability. The third man failed of great repute, owing chiefly to his advanced—rather too much advanced—visionings. He wanted to talk across the ocean by telephone at a time when the cable interests successfully prevented him from commercializing his apparatus. And he died a disappointed inventor. But he had the stuff in him, the thing that makes for human greatness, just as had the other and more successful two men with whom I as a designer was privileged to work. All were men of kindly spirit, of broad outlook, of unselfish devotion to worldly interests. Each was a humanitarian. Each saw things as they are, and each saw things as they should be, and each thought much on problems of human welfare and betterment. Of such men in civic affairs the nation, and indeed the entire world of nations, has had but a sad too few in the past. It is to be hoped, and it is the belief of the writer, that engineers will become more plentiful in civic life in the future.

I have always believed that the man who reached an advanced age without a sizable bank-account is a fact which would well serve as a definition as to what constitutes an idealist. There are many such men—meaning, of course, men having a level set of brains, and not mental incompetents. Such men are inclined to things other than the accumulation of bank-accounts. They strive toward goals which to them are more worth while—self-improvement, for instance, spiritual growth being a better term. Of such men were the world's acknowledged saviors. A man who can wilfully thrust oars against the current of a stream flowing currency-wise, in such a way as to force himself into a back eddy or pool more or less stagnant, is a man pronouncedly great among men. The world is loath to recognize such a man for what he is; yet such men have lived and still live and will continue to live, always more for others than for themselves—seeing life in the true, in other and more gracious words.

61

Engineers, in the abstract, are such men. The accumulation of money is secondary with them. Their work holds first place in importance. Possessed of that professional pride which will not permit a man to set aside his work and enter a more lucrative and materially satisfactory field of endeavor—if he starve in his obstinacy—engineers are men of the temperament, aside from the training, to minister to public needs and desires. Self-effacement is the engineer's chief characteristic. He views largely and without bias. He can see things from the other fellow's angle because he is not an engineer if he has not the gift of imagination. The successful engineer has this most precious of endowments, and, having it, cannot but be possessed also of kindliness and sympathy, which are imagination's own brothers. Kindliness and sympathy are needed in the high places of our government for the people by the people. And because men in time gravitate to their rightful sphere of usefulness through the workings of an all-wise Providence, engineers already have turned and are turning toward the administration of public affairs.

62

IX

CODE OF ETHICS

All engineering societies have a code of ethics for the guidance of their membership bodies. In each case it is a code based upon other and older codes, codes long in practice among professional men, such as lawyers and doctors. It is a code built up on Christian principles, as it should be, and rarely is it ignored among men of the profession. To do unto others as you would have others do unto you is the basis of its precepts, though more concretely it aims to guide the engineer in his business intercourse with other men in such a way as to give all an equal chance without transgressing the law. The so-called building codes in effect in large cities are intended to hold engineers to restrictions for the greatest good of the greatest number, and the code of ethics in practice among each of the engineering professions likewise was devised toward this end. There seems to be need for it.

Perhaps by pointing out where engineers sometimes transgress, the writer more effectively can indicate the need of a code and the principles of which the engineering code of ethics consists. Even to-day there are engineers digressing from the path indicated by the professional body, though in such a way as to benefit still by the protection of the law, and to be not openly susceptible to admonition from the engineering societies' committees. Engineers of this stamp at best are but tricksters. Actually, they should be debarred from practice, just as the legal fraternity takes effective action against members of the bar who go outside the pale, though nothing is ever done to engineers. Engineering organizations in this regard are weak. The man's name should at least be posted, or, better still, published in the society's bulletin, so that the fraternity at large could know, and, knowing, could warn men with capital to invest—the trickster's especial prey—for its own welfare.

There was an engineer brought to the attention of the writer whose activities were devoted to securing for his clients men of no mechanical knowledge who yet wanted something done by machinery. A manufacturer of paper dolls, say, having entered upon this phase of manufacturing only because he had money to invest and not because he was interested in mechanics, would see the need in his plant for additional mechanical devices to cut down manufacturing costs. The engineer to whom I have reference would find this type of manufacturer his particular "meat," because of the man's ignorance of mechanics, and, after clinching him with a contract drawn up by the engineer's lawyer, would undertake to devise for this manufacturer a perpetual-motion machine, if that happened to be what the manufacturer wanted. The engineer conducted a machine-shop in connection with his "consulting" office, where, at a dollar an hour for the use of his machine-tools, he would "develop" his ideas, as passed upon by the manufacturer who knew no more of construction or the reading of mechanical drawings than he did of the chicanery of the engineer, and in this way roll up the costs against the unfortunate. In the end the engineer might and might not produce a satisfactory working machine. There was nothing in the contract about this—save only as it protected the engineer. What was indeed produced was a list of costs for the development often of several designs of a given idea that to say the least were heartrending.

65

Then there is the engineer who for a consideration will bear false testimony against his neighbor, or his neighbor's ox. This happens most frequently in municipal traction or lighting wars, set before tribunals under the caption of "The People *vs.* the S. S. Street Railway Company," or in a battle of alleged infringement of patent rights. There are engineering experts, just as there are legal experts, who deem it within their code of ethics to address themselves and their energies toward the refutation of such claims, however wrong or right these claims may be. Engineering is an exact science. It is based on principles hardly refutable. Yet there are engineers who will and can confound these principles before a court of law in such manner as to win for their clients a decision of non-suit where the facts point glaringly to infringement—in the matter of mechanics—or to win for their clients a favorable decision in the matter of costs of maintenance and operation of a railway, in a case of this kind. As has been said, figures don't lie, but figurers sometimes do.

Other instances of breach of engineering ethics, however otherwise secure from the clutches of the law, occur to the writer, but the two just cited ought to serve. At best, 66 the topic is unpleasant and by no means indicates the character of the profession as a whole. Where there is one engineer who will perjure himself in the fashion as set forth above there are many thousands of engineers who could not be bought for this purpose at any amount of money. The profession of engineering is notably clean; its code of ethics rigidly adhered to; the rights of others, both in and out of the profession, regarded with something akin to sacredness. Engineers, as a body, for instance, possess a peculiarly rigid idea concerning themselves in relation to branches of the profession outside their own and yet intimately close to their own. Called in as an expert in the matter of heating and lighting a building, say, the heating and lighting engineer will rigidly confine himself to this phase of the engineering venture and to no other, however he may find his work again and again overlapping the work of the structural engineer or the industrial engineer—phases concerning which he may possess important knowledge. He regards these things as strictly none of his business, and in doing so conserves the esteem and friendship of his confrères.

The code of ethics is a liberal one among the engineering groups. It has been laid down 67 with an eye to fairness both for the practitioner and the client. Rigidly held to, it will admit of no engineer going far wrong in the practice of his profession, and, broken, will not land him in jail. It is presupposed that engineers are men of intelligence. A man of intelligence will hold himself to the spirit of the Ten Commandments if he would attain to success, and to the letter of them if he would be happy during the declining days of his life. Most engineers realize this and accept it as their every-day working creed. Life to them, like the medium through which they give expression to their ideas, is a matter of mathematics. Two steps taken in a wrong direction mean an equal number of steps forcibly retraced—or the whole problem goes wrong. Engineers rarely take the two steps in the wrong direction. When they do take wrong steps they are quick to right them. For the code is always before their eyes.

68

X

FUTURE OF THE ENGINEER

Just at present the future of the engineer is more richly promising than it might otherwise have been but for the war. Due to the period of reconstruction now confronting the world, a work almost wholly that of the engineering professions, engineers for a period of a decade at least are destined to be overburdened with projects. Nor will any one branch be occupied to the exclusion of any other branch or branches. Civil and structural engineers will, as a matter of course, have the first call; but with the work of these men well under way — consisting of the reconstruction of towns and cities — mechanical and electrical men will necessarily be called upon, with, no doubt, liberal demand for mining engineers. Each branch will have its place and serve its usefulness in the order as the reconstruction work itself will fall, with the result that all branches of the profession will be busily occupied.

69

Manufacturers have been ready or are getting ready for this unprecedented promised activity for some little time. Representatives are flocking abroad on every boat sailing from these shores with schemes and plans for the rapid upbuilding of devastated Europe. These men, for the most part, are engineers embracing all branches of the profession, and each is a man especially well qualified to serve in his branch. In a way he is a specialist. He may represent a giant structural organization, or a machine-tool manufacturer, or an electric-lighting and power concern — any one of the many fields of industrial enterprises whose product is needed to place demoralized France and Belgium back upon a productive basis. For when the construction period is over with there will be need for machine-tools and equipment for operating these tools, such as engines and boilers and motors, all of which come properly under the head of engineering productive enterprises.

Engineers — especially American engineers — will be in great demand, as they are already. Nor will the close of the reconstruction period witness an abatement of this demand. Having once entered the foreign field on a large scale, they will of necessity 70 continue to be in demand not only for the furtherance of industrial projects, but for purposes of maintaining that which has been installed at their hands. Machinery has a way of needing periodical overhauling—even the best of machinery—and this will entail the services of many engineers for long after the machinery itself has been set up. The services of erecting engines, operating engineers, supervising engineers—known more properly as industrial engineers—following, as the need will, close upon the heels of the constructing and selling men—will keep the many branches alive and in foreign trade for much more than a decade—or so it seems to the writer. Other nations may, of course, whip into the field and in time crowd out the more distant—meaning American—engineers and engineering products. But I don't think so, because of the acknowledged supremacy of American engineers in many directions. The war itself taught the world that we possessed such a supremacy, and the world will be slow to forget—especially the purchasing side of nations themselves so crippled of man-power as to be for a generation well-nigh helpless.

So the immediate future of the engineer is richly promising. It is so rich with promise 71 that a young man could hardly do better than to enter upon engineering as a life-work, provided he has no particular choice of careers, and would enter upon an attractive and scopeful one. His work is already laid out for him. Taking up a course of study leading to the degree of M.E., or C.E., or E.E., in four years, upon graduating, he can retrace his way, or the way of his brother, over the battle-fields of Europe, a constructive rather than a destructive agent now, a torch-bearer, a pilgrim, a son of democracy once again advancing the standard in the interests of humanity. He may do this as a mechanical engineer, as a civil engineer, as an electrical engineer, as a mining engineer; it matters not. What does matter is that he will be carrying Old Glory, in spirit if not in the letter, to the distant outposts—the especial province of the Anglo-Saxon race, anyway, from the beginnings of this race—and so serving to maintain the respect and affection already established in these countries by our soldiery. To the writer the thing looks mighty attractive.

Yet the young engineer's future need not lie in distant places necessarily. He may stay at home and still have his work cut out for him. The promised unparalleled activity 72 in the field of engineering on the other side cannot but enlarge and accentuate the activity on this side of the water. Plants will be operating full blast to catch up with the demand imposed by this abnormal activity, and thus the engineer will perforce bear the burdens of production. He will bear them in all directions, since industrial activity means engineering activity, and the work of production cannot go on without him. In the mines, the mills, the quarries, the foundry, the machine-shop, the pattern-shop, the drafting-room, the engineering offices, the consulting divisions—all these, necessitating as they do the employment of one or more engineers in at least a supervising capacity, will have urgent need for his services. Constructive work always, he will grow as his work grows, and because the growth of his work under these abnormal conditions will be of itself abnormal, his own growth under these conditions will be abnormal. He will find himself a full engineer before his rightful time.

Right here it would be well to point out to the young graduate the importance of getting under a capable engineer. For, much as the writer dislikes to admit it, there are engineers who are not capable and who yet occupy positions of great responsibility. The 73 young engineer, fresh from college and a bit puzzled as to the game as a whole, if he accept a connection under an engineer, for instance, whose inventive ideas are impractical, will unwittingly absorb such a man's viewpoint on construction, and so spoil himself as an engineer for all time to come. Cases like this are not rare. The writer personally knows of more than one young man who enlisted under an engineer whose ideas on administration probably accounted, being as they were good ideas, for his position of authority over matters not strictly of an administrative nature. The man wanted to exercise his authority over all things within his department—not the least of which was machine design—with the result that the young graduate's normally practical viewpoint on matters of construction became warped into that of the man over him, and continued warped for so long as he remained under this man, and frequently longer, indeed, to the end of his engineering career. The young engineer must pick his boss as our young men are facetiously advised to pick their parents. The wrong selection will prove disastrous to him in after-life.

Which is but an aside—though a very important one. To emulate a weakling in 74 whatever walk of life, be it painting or writing or engineering, means to begin wrong. Everybody knows the importance of a right beginning. It is no less true of the young engineer than of others.

And what with the example set by Herbert Hoover and other dollar-a-year men, mostly engineers, in the nation's administrative affairs during the war, the future of the engineer looks bright in these quarters as well as in quarters embracing engineering constructive work wholly. The engineer of the future undoubtedly will take active part in municipal and national affairs, more likely than not in time entering upon a political career as a side interest, as the lawyer enters upon it to-day, within time—so it seems to the writer—members of the engineering professions occupying positions of great trust, such as state governorships and—who knows?—the Presidency itself. Certainly the hand points this way. More and more engineers are coming into prominence in the public eye, and with every member of the profession so coming, the respect for men of his profession multiplies among laymen. It is not too much to say, therefore, that engineers are destined to fill places of great political power. It is to be hoped that they 75 are. Whether they do or not, the future at this writing amply promises it, and so forcibly that it may well be included as existing for the engineer, as being a part of the future of the engineer.

76

XI

WHAT CONSTITUTES ENGINEERING SUCCESS

A graduate of Cornell, in the class of '05, after placing away his diploma where it could not trouble him through suggestiveness, accepted a position with a large manufacturing concern in western Pennsylvania. He was twenty-three years old. He went into the shop to get the practical side of certain theories imposed upon his receptive nature through four long years of study in a mechanical-engineering course. The concern manufactured among other things steam-turbines, and this young man, having demonstrated in school his particular aptitude for thermodynamics—the study of heat and its units in its application to engines, and the like—entered the erecting department. Donning overalls, and with ordinary rule in his hip pocket—as against the slide-rule with which he had worked out his theoretical calculations during his college 77 years—he went to work at whatever was assigned him as a task by his superiors—shop foremen, assistant superintendent, occasionally an engineer from the office.

This young man did many things. He helped to assemble turbine parts; carried word of petty alterations to the proper officials: assisted in the work of making tests; made detailed reports on the machine's performance; screwed up and backed off nuts; in short, got very well acquainted with the steam-turbine as manufactured by this company. He knew the fundamentals of machine construction, and an understanding of the details of this particular type of turbine therefore came easy to him. He worked shop hours, carried his lunch in a box, changed his overalls every Monday like a veteran. Usually his overalls more than needed changing, because he was not afraid of the grease and grime with which he came into contact throughout the day. He liked the work and went to it like a dog to a bone. He was applying in a practical way what he had learned in college of a theoretical nature, and finding the thing of amazing interest.

He made progress. In time his work was brought to the attention of the chief engineer, and one day, when the president of the 78 company, who was also an inventor of national repute and responsible for the design of the turbine being manufactured by the organization, wanted to make certain bold changes in the design, the chief engineer sent for the young engineer whose work in college in thermodynamics had won for him certain honors, with the result that our hero found himself presently seated opposite the president at a table in the latter's office, engaged in working out calculations on his slide-rule—calculations beyond the powers of the president, because he was not a heavy theoretician. This call was a big advance indeed, for it marked him as a man of promise—a "comer"—in the concern. The president liked the ease with which the young engineer "got" him in the matter of the proposed changes, and quite before either realized it both were talking freely, exchanging ideas, in the field of turbine construction generally. The young man unconsciously was driving home the fact that he was a capable engineer, one who, while still lacking in broad experience, was nevertheless possessed of the proper attitude toward engineering as a whole to compel the interest and attention of his superior.

The young man eventually was sent out upon the road as an erecting man. In this 79 work he discovered certain operating faults in the design, and, reporting these faults to the home office, observed that not a few were remedied in subsequent designs. He moved about the country from place to place, setting up and operating steam-turbines, until there came the blissful day when he was called back to join the engineering staff in work covering design. Laying aside his overalls, he emerged as a crisp young engineer in a linen collar and nifty cravat—although not till later did he don a cream-colored waistcoat—and thereafter his hours were seven instead of nine. With a desk and a stenographer he entered upon work of a somewhat statistical character. He followed the designs of rival companies as best he could through their advertising and articles covering their respective designs appearing in the technical journals, and about this time also applied for admission, and was granted it, in the foremost engineering society embracing his particular branch of the profession. He was still making progress.

Likewise, he was rapidly becoming an expert in the field of steam-turbines. His work in the shop, together with his experience on the road, both as an erecting man and operating engineer, had eminently fitted him for 80 valuable service in the home office as an engineer overseeing design. His work in charge of design, where his knowledge of what had given service both good and bad in details of construction while he was in the field, was extremely valuable to the designer himself, was rapidly rounding him out as a steam-turbine man. His salary had gone up apace with his progress; he had met the right girl at a club dance in the suburban town where he had taken modest quarters; he was rapidly headed toward success both as an engineer and a citizen. He had been out of school probably six years, and was still a very young man, with all the world practically before him.

One day he was asked by the chief engineer of the concern to journey to New York, and read a paper before his engineering society at one of the regular annual meetings, on the subject of thermodynamics in its relation to the company's own product—the turbine. He tipped over his chair in his eagerness to get out of the office and on the train. He realized the importance of this opportunity. He was to appear before his fellow-engineers—the best and most capable and prominent in the profession—and to appear as an authority on his subject! The thing 81 was another step forward. He prepared a paper, basing it on his six years' experience in steam-turbines, and when he reached New York had something of value to tell his brother engineers. The meeting was held in the afternoon, and, dressing for the part, he stepped out upon the platform before a gathering of some eight or nine hundred engineers and delivered himself of his subject with credit to himself and to his organization. Not only that. In the rebuttal, when engineers seated in the auditorium rose to confound him with questions—engineers representing rival turbine concerns—he proved himself quick at the bat and more than once confounded those who would confound him.

He was making his mark on the industrial times. His paper was reviewed in the technical journals and almost overnight our young hero found himself recognized as an authority in his chosen branch. He was sought out for other articles by technical editors, his associates in the home plant generously commended him for his work; his salary received another elevation; he called on the girl that night and had her set the date. Then he plugged for salvation—further knowledge as a turbine man—harder than 82 ever. Having won the full confidence of the officials of the company by this time, he was

53

given free voice in all matters having to do with the design of their product, and shortly after his first little boy was born was promoted to the position of assistant chief engineer. He served in this capacity for two years, and then, realizing that he had gone as far up in the organization as it was physically possible to go, owing to the fact that the chief engineer was the president's sister's husband — or something like that — he accepted an offer from one of the rival concerns manufacturing turbines and entered the organization as chief engineer at a salary too big to mention. Our young friend had at last arrived.

Yet his success was not quite complete, nor will it be complete, until he sets up, as he assuredly will some day, as a consulting engineer. When he at last does this, when he swings out his shingle to the breeze, he will then have attained to the maximum of possible success as an engineer. Already recognized as being possessed of a fine discrimination in matters of engineering moment, especially in thermodynamics as related to turbines, he has but gone up in channels early laid out for him, and indicated to him, in 83 his college days. His direction even then was clearly marked. All he had to do, and all he did do, was to develop himself in this single direction. He did nothing that would be impossible to any other engineering graduate. Merely he hewed to the line — persisted in remaining in the one branch of the game — met with his reward in time just as any young man would meet with it. There was nothing of phenomenal character, nothing of the genius, revealed in what he did. His way is open to all. And it is a way both worthy and admirable, for to-day this engineer stands high in his profession and is meeting with financial reward in keeping with his position among engineers.

There you have in the tracing of one engineer's progress to success precisely what constitutes engineering success. The details may differ, but the principles and the rewards will be the same, whether you enter upon civil or mechanical or mining or electrical engineering. Success in engineering constitutes certain satisfactory money rewards and an even more satisfactory recognition by one's associates and fellows. Success in anything is that. A man must work for them, however. There never was and never will be a rainbow path to the heights. Toil and an abiding 84 faith in one's own capabilities—these make for success. Success makes for happiness, and happiness, as everybody knows, is all there is to this life.

I wish all men happiness.

85

XII

THE PERSONAL SIDE

As to the personal side of engineering as a career, if it would be a source of gratification to you to know that you were helping to build up the civilized world, then you should enter the engineering profession. Because men differ in their ideas as to what constitutes a full life—some placing ideal homes above all things, some seeking continuously diversified sources of pleasure, some wanting nothing better than a fine library or freedom to cultivate taste in pictures, some wishing only to surround themselves with interesting people, some wanting nothing but an accumulation of dollars, some wishing but for power of control over others—all men would not find the full life in engineering. Yet the majority of men would, because the profession holds that which would appeal to a great many different ideas as to what a complete life consists of. Engineering as a profession is scientific, idealistic, constructive, 86 profitable. It is combative—in the sense that it shapes nature's forces—and it calls for a sense of artistry in its practitioners. Added to these, it embraces a certain kind of profound knowledge the possession of which is always a source of pride to the owner.

Let me explain this last. The engineer, being as he is a man who views things objectively, notes details in everything that comes under his eye, be it dwelling or automobile, or bookbinding or highway. The layman does not. The layman, outside his work, sees only the thing itself, when looking at it—the general outline. But the engineer, trained to note details in construction, observes detail at a glance, and does it almost subconsciously, if not immediately after leaving school, then assuredly later, after he has been practicing his profession for a time. His outlook is objectively critical. Entering a house for the first time, and trained as a mechanical engineer, he will note the character of the woodwork, the decorations, the atmosphere, the arrangement of the furnishings, all with the same facility that he will note details upon entering for the first time a power-station or a manufacturing plant—things within his own province.

Nor is this faculty confined to the concrete. 87 Engineers are of that deeply instinctive race of folk who perceive cause in effect with the lightning swiftness of a wild animal. If they are not this when entering upon the profession, assuredly they become so after a period spent in the work. Something about the practice of engineering breeds it—breeds this objective seeing and abstract reasoning—and to be possessed of it is to get more out of life than otherwise is possible. Which possibly accounts for the fact that engineers as a group seem to have a common-sense viewpoint of things, one that is frankly acknowledged and drawn upon when needed by men in other walks of life. Engineers are extremely practical-minded, and this makes for a certain outlook that will not permit of visionary scaring away from the common sense and the practical on the part of its possessor. Engineers know why things occur without having witnessed even the occurrence itself. Their powers of reasoning are developed to degrees beyond the average—or they seem to be—and out of this comes one of the sources of gratification on the personal side to the man who pursues engineering as a profession.

The thing spreads out as I contemplate it. I would make so bold as to say that the man 88 of engineering training will see more at a glance when first viewing the Grand Cañon, say, than will any other professionally trained man. Should the Cañon collapse, he would know instantly why it collapsed. He could give an opinion on the wonderful color effects that would interest the artist, and he would know without hesitation how best to descend to the bottom and wherein to seek the easiest trail. All this, without his being a civil or a mining engineer, understand; merely a man trained in constructive mechanics. On the other hand, the mining or the civil man would view the wreckage of a locomotive accident and see in the debris, select from the snarl of tangled wheels and driving-arms and axles a ready picture of the nature of the accident and how much of the wreckage offered possibilities for repair. Again, the engineer sees in a tree, with its tapering trunk, the symbol of all tower construction, just as he sees in the shape of a man's arm the pattern to follow when devising a cast-iron lever for an automatic machine. He sees things, does the engineer; sees objectively; follows nature throughout.

All this being true, the engineer has a rather interesting life of it. For not only does he see a little more clearly than otherwise would be possible to him without his education and training, but also he does things with his hands that come easy to him without previously having undertaken them. The engineer can do much around his own home, if he so choose, that of itself is a source of great satisfaction. Engineers can swing doors, build fireplaces, landscape, erect fences, make garden, and can perform these tasks with a degree of neatness and skill that brings favorable comment from journeymen whose vocations this work is, and do the work without training whatsoever in the work. Wall-papering, painting, carpentering, laying up of brick, or the placing of a dry wall—plastering, glazing—the list is endless that as side-plays are possible to the man with an engineering training. He need not do these things, ever; but if he wants ever to do them, he finds that he can do them and do a creditable job of each, and this without his ever having turned his hand to the work before.

Which sums up in a measure the personal side. The engineer is not a superior being. Merely he is a man possessed of a highly specialized education and training which peculiarly fits him for any practical work, and out of this work, for practical thinking of the kind known as constructive. Being constructive with his hands, he cannot but in time become constructive with his brain. Being constructive as a thinker first, he cannot but become constructive as a doer later. The one hinges closely on the other, and having both, as the engineer must who would be a successful engineer, he has as much of the world under his control as comes to any man, and, in a great degree, more than is the favorable lot of most men. For the engineer is both a thinker and a doer. Ponder that—you. Men are either one or the other—most men—and rarely are they both. Either side of their brain has been developed at the expense of the other side. Not so with the engineer. The successful engineer is both thinker and doer—must be in his profession. It seems to me that engineering has many beautiful attractions as a profession.